MATTER, ENERGY AND MENTALITY
EXPLORING METAPHYSICAL REALITY

Matter, Energy and Mentality
Exploring Metaphysical Reality

RICHARD RYDON

Speculative Non-fiction

Copyright © 2013 by Richard Rydon

All Rights Reserved.

No part of this book may be reprinted or reproduced or utilised by electronic, mechanical or other means including photocopying or any information storage or retrieval system without permission in writing from the publishers.

US Trade Edition: November 2013

ISBN: 978-1-291-58403-5

Cover image: Photo 8993154 © www.123RF.com
Image credit: Kittikun Atsawintarangkul, fine artist

Published by www.lulu.com

He, who thro' vast immensity can pierce,
See worlds on worlds compose one universe,
Observe how system into system runs,
What other planets circle other suns,
What vary'd being peoples ev'ry star,
May tell why Heav'n has made us as we are.

— Alexander Pope
Essay on Man, 1733

Contents

Title Page	iii
Copyright	iv
Quotation	v
Chapter 1. A Metaphysical Frame of Reference	1
1.1. Sectors of Reality	1
1.2. Things Change	2
Chapter 2. The Threefold Structure of Reality	4
2.1. Reality	
2.2. Matter	4
2.3. Energy	4
2.4. Mentality	5
2.5. Things	6
2.6. Living Systems	6
2.6.1. Organization	6
2.6.2. Evolution	8
2.6.3. Coordination	11
2.7. Conclusion	12
Chapter 3. Organism — The Source of Unnecessary Action	13
3.1. Living Systems	13
3.2. Treatment	14
Chapter 4. Information and Entropy	17
4.1. Events and Observations	17
4.2. Information	17
4.3. More About Information	19
4.4. More About Entropy	20
Chapter 5. Three Lesser Laws of Reality	23
5.1. More about Energy Redistribution	23
5.2. The First Lesser Law of Reality — Equation 1	23
5.3. The Second Lesser Law of Reality — Equation 2	24
5.3.1. Maximum Capacity and Relative Size	24
5.3.2. Maximum Capacity and Organisational Complexity	25
5.4. The Third Lesser Law of Reality — Equation 3	27
5.5. Summary	29
5.6. Conclusion	30

Appendix 1. Logarithms, Summations and Limits 31
   A.1.1. Logarithms 31
   A.1.2. Summations 32
   A.1.3. Limits 32

Appendix 2. Information and Observation 33
   A.2.1. Observation of a System that Always Changes State 33
   A.2.2. Observation of a System that Sometimes Changes State 35

Appendix 3. More About Functional Information 36
   A.3.1. Biological Functional Information Distributions 36
   A.3.2. Active Transport: A Biophysical Example 36

Appendix 4. Different Levels of Organisation 39
   A.4.1. Organisation 39
   A.4.2. Abstract Organisation 39
   A.4.3. Physical Organisation: Machines 40
   A.4.4. Biological Organisation: Cells 41
      A.4.4.1. Genetic Information 42
      A.4.4.2. Macromolecules 43
      A.4.4.3. Steady State Systems 43
      A.4.4.4. A Worked Example: Active Transport Again 44
   A.4.5. And Finally ... 46
      A.4.5.1. Cellular Information Processing 46
      A.4.5.2. Ability to Codify Information 46

Bibliography 48
About the Author 53

Chapter 1
A Metaphysical Frame of Reference

1.1. Sectors of Reality

Once a formula is stated, it can be followed by an explanation of what each of its terms mean. Of course, this in no way proves that the formula is correct but it certainly focuses attention on the essential points to be considered. It is one of the quickest and clearest ways of communicating; it provides a frame of reference and the arguments thereof.

The following, then, are the two metaphysical propositions on which this book is built:

> There are three, fundamental, necessary and sufficient but also essentially independent, indefinable Sectors of Reality; namely, Matter, Energy and Mentality, from which all things arise.

> The Universe is the field of all things contingent on the co-extensiveness of Matter, Energy and Mentality, and can only be ultimately accounted for in terms of them.

In Set Theory, we learn to illustrate by means of Diagrams, all the individuals included in a given class. According to Bodkin (1963),

> "With this understanding we can easily represent by diagrams to the eye what 'precisely' is asserted in each proposition and so make the matter clearer to the senses."

The relationships between the terms Nature, Reality and Universe, as well as Matter, Energy and Mentality, are represented in such a way in Figure 1.1.

Thus, as shown in the Diagram, Nature is a broader term than Universe, and includes not only co-extensive regions of Matter, Energy and Mentality but also all the regions containing them. And by the same token, Reality is an even broader term than Nature, and includes other regions of which we can have little or no knowledge. However, for all practical purposes, the three terms Reality, Nature and Universe, may be used interchangeably.

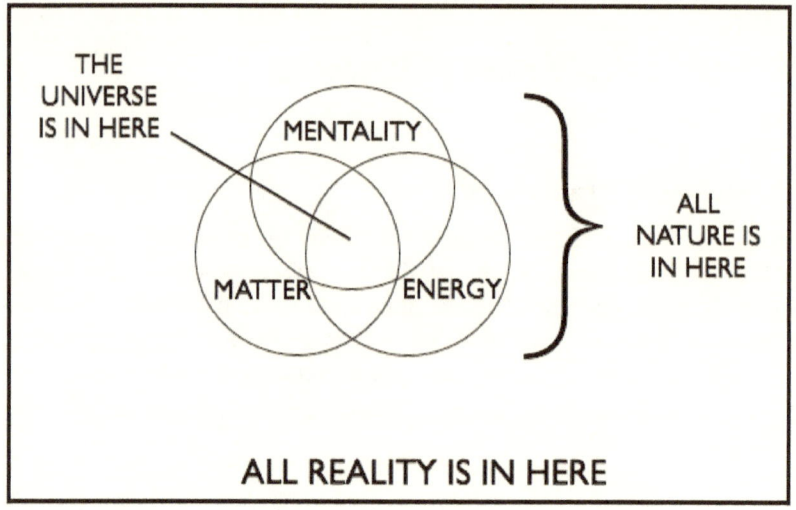

Figure 1.1. The Universe, Nature and Reality.

Again, it can be seen that the terms Matter, Energy and Mentality are independent, and are not contained within each other. They are absolutely distinct entities in themselves. They are indefinable only because they require no explanation.

Bodkin (1963) holds:

"... if a thing has no parts you cannot enumerate its parts."

Nevertheless, they can be as fully understood as other indefinables, such as the perception of colour or taste. And, because they are simple, they are called fundamental or irreducible elements of Nature. The description, necessary and sufficient, can also be ascribed to them in the sense that the Universe is dependent on them.

1.2. Things Change

A system may be any group of related phenomena, the exact number (all of which may not be known) constituting the Quality of the System, and the exact frequency of each one of them constituting the Quantity of the System. Every such system will consist of at least one phenomenon which can occur at least once. In practice, the Quality and Quantity of a System may be impossible to determine.

From the simple observation that Things Change, the following can be concluded:

There is a *discontinuous* element capable of assuming various states — Matter.

There is an *active* element capable of bringing about changes of state — Energy.

There is a *storage* element capable of surviving changes of state — Mentality.

So, the terms Matter, Energy and Mentality, are not only qualitative but also quantitative. In other words, relationships between them are possible. Furthermore, in the deepest sense, they are not independently isolatable entities. To speak of Matter, Energy or Mentality in isolation in the Universe is meaningless; they can only be considered in relation to each other. They are the lenses through which everything in the Universe exists and can be viewed.

Chapter 2
The Threefold Structure of Reality

2.1. Reality

Reality is comprised of the three fundamental Sectors:

>Matter (Realm of Probability).

>Energy (Realm of Possibility).

>Mentality (Realm of Purpose).

These three fundamental Sectors of Reality constitute the lowest levels or degrees of Existence and are co-extensive in the Universe.

2.2. Matter

This Sector of Reality comprises all the purely material aspects of the Universe throughout the whole of space and time. Everything concerned with Matter in all its forms and states is contained in this Sector.

Because of its particulate nature, the material Sector of Reality is, in the limit, governed mainly by the laws of Probability; although, when viewed from a distance, its behaviour can be predicted with amazing accuracy and certainty. However, we will not be concerned with any specialised study of matter, as for example that of the chemist or physicist, but rather with the essence of Matter, which is a substratum capable of rearrangement to give rise to numerous definable Things.

2.3. Energy

Energy may be considered as the interface between Matter and Mentality, bringing them together. Changes occur when Energy is redistributed within the imposed limitations of Matter and Mentality.

There are two aspects of Energy required to describe how the three Sectors of Reality—Matter, Energy and Mentality—interact in Living Systems:

>Aspect 1. This is the familiar Physical Energy, identical in every way with the Energy of the Physicist or Chemist.

>Aspect 2. This is not so familiar. Virtual Energy is related to Energy Redistribution, and whether it is necessary or unnecessary.

An analogy from physics helps to explain this distinction.

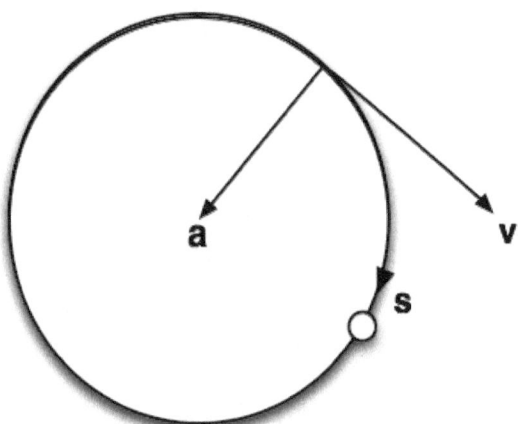

Figure 2.1. Uniform circular motion.

In Figure 2.1, the acceleration (a) towards the center of the circle keeps changing the velocity (v) of the object (O) resulting in a circular motion with constant speed (s). So, an object travelling at a constant speed (s) still requires a constant acceleration (a) to enable it to describe a circle. However, it is not necessary for an object to travel in a circular. The same amount of Energy can be expended by travelling in a straight line. Similarly, Living Systems, while not requiring more Physical Energy than inanimate forms separately performing the same functions, yet, require Energy Redistribution to maintain themselves.

2.4. Mentality

Mentality constitutes another Sector of Realty. It is important to distinguish between Mentality and its attributes, and it may be of assistance to visualise Mentality as an entity akin to Matter and Energy, insofar as it is as widely distributed throughout the Universe as they are. Also, since the word Mind refers to a Localised Mental Activity, an important distinction can be made between Mentality and Mind. The former is universal, simple and independent of Living Systems and the latter is localised, complex and dependent on the presence of Living Systems. Mind enables the characteristics of Mentality to be fully expressed.

2.5. Things

Again since the Universe is the field of all things contingent on and definable in terms of Matter, Energy and Mentality, in this sense Body and Mind are things, but they are not capable of existing independently. They are simultaneous. However, it is more convenient to reserve the term Thing for non-simultaneous contingents, and to refer to Body and Mind as the two simultaneous aspects of all Things. In other words, Things are mutually exclusive entities containing two mutually inclusive aspects Body and Mind. But Things overlap and can never be mutually exclusive in any absolute sense. So, mutually exclusive means that $Thing_1$ could exist in the absence of $Thing_2$.

It may be noted that this merely removes the problem of Body and Mind back a stage further to Matter, Energy and Mentality. This is true; but, it has the advantage of making the concepts of Body and Mind complex (which they are) and specifiable in terms of the more fundamental entities of Matter, Energy and Mentality.

2.6. Living Systems

Living Systems have three aspects:

>Organisation (Spatial Aspect).

>Evolution (Temporal Aspect).

>Coordination (Mental Aspect).

These three aspects will be illustrated diagrammatically in the following sections.

2.6.1. Organization

It is necessary here to introduce another term. Everything that exists may be given a Number of Definition to indicate its level of existence in Reality. When considering Living Systems, the Number of Definition is a useful Concept. Each Organism may in theory be assigned one.

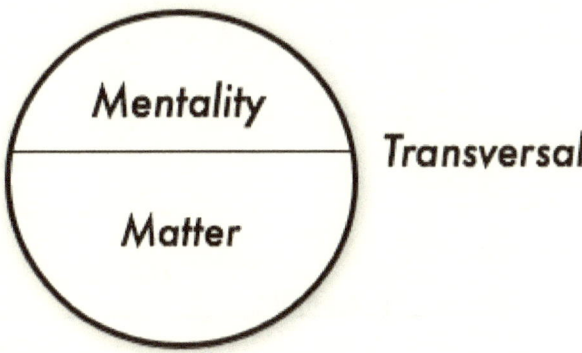

Figure 2.2. The diagrammatic organism.

In the diagrammatic representation of an Organism in Figure 2.2, the line separating Matter from Mentality may be called the Transversal. The position of the Transversal is determined by the Number of Definition, which depends on the type of System in question.

The Number of Definition is defined as the logarithm of the ratio of Mentality to Matter in the diagrammatic representation of an Organism. The mathematical device of taking logarithms is used here, because of its great advantage in representing very wide variations in a more manageable scale, particularly in graphs. No knowledge of logarithms is necessary to understand the graphs in this Chapter. However, in the next Chapter, some basic mathematics is used.

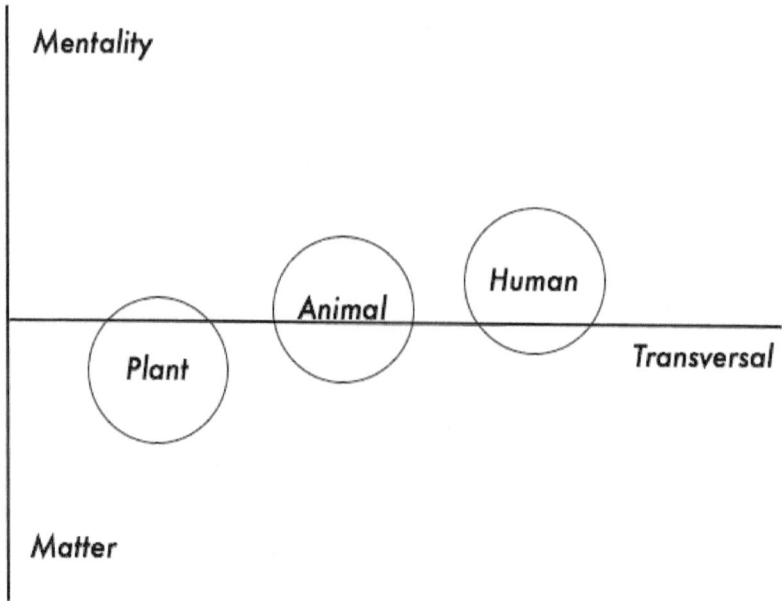

Figure 2.3. Some typical organisms.

Figure 2.3 illustrates, in pictorial form, some typical Organisms as circles. The regions of Mentality and Matter are shown above and below a horizontal line—the Transversal—crossing through each Organism. The Number of Definition locates the Organism with respect to the Transversal. For example, this could be −7 for a typical Plant, +1 for a typical Animal and +5 for a typical Human. The numbers here are arbitrary. In general, the higher the Number of Definition, the more complex the Organism.

2.6.2. Evolution

Consider a quasi-independent Lump of Reality and make this Lump travel rapidly through the aeons of time that the Earth has existed. Figure 2.4 is a graph of the apparent path of this Lump with respect to the two regions of Matter and Mentality previously defined.

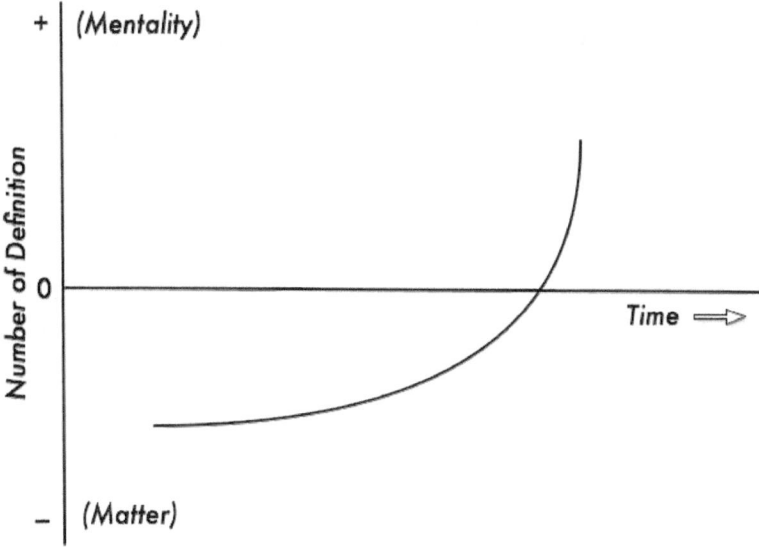

Figure 2.4. Evolution, Graph 1.

Observe a certain tendency—that is what evolution is all about. But, there is a fundamental blunder. We assumed that the Lump passed directly and successively into different systems of organisation as if growing or evolving continuously. Thus, following this to its logical conclusion, Matter eventually turns into Mentality. Surely this is not so. Perhaps the illusion depends simply on the illustration. Let us take another illustration.

Let us take the average Number of Definition of all such Lumps of Reality, in existence at any particular time, and plot this against a rapidly moving chronological background, as illustrated in Figure 2.5.

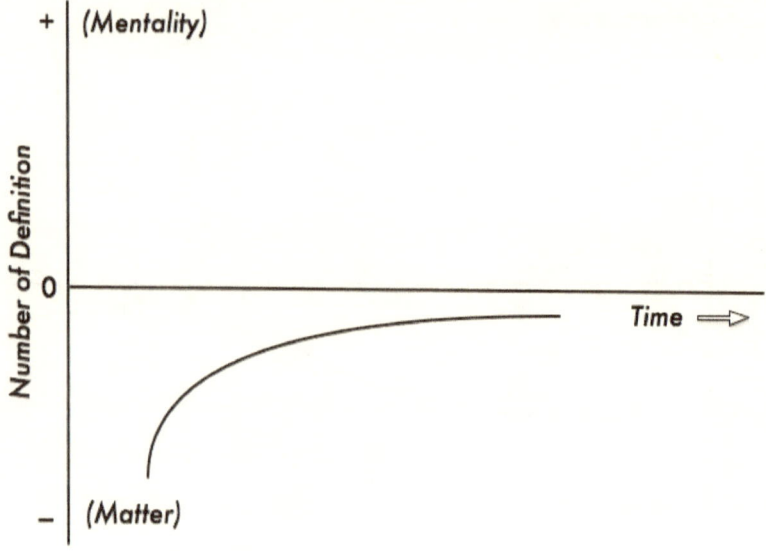

Figure 2.5. Evolution, Graph 2.

In this case, we see that the asymptote of the curve is the Transversal of Reality as defined above. I am thus led to formulate a more refined definition of Evolution:

> Evolution is that phenomenon which tends to bring together two Sectors of Reality, Matter and Mentality, through the intervention of a third Sector, namely, Energy. And, having done so, will maintain the three Sectors in Natural Equilibrium.

Thus, through the Phenomenon of Life, the Transversal of Reality is gradually becoming less of a Boundary and more of a Focus.

In some Organisms the Number of Definition remains constant during the life span, and even over much longer periods of time, if this does not cause a new organisation or species to arise. In other Organisms, the Number of Definition varies more rapidly, repeating its variation within the life span of each new individual of the species. Humans fall into this category.

The next illustration depicts a typical Human, where the Number of Definition is plotted on the vertical axis, as before, and the human's development on the other axis. In Figure 2.6, the graph starts at the moment of conception. At this stage, the Number of Definition is that of a typical cell, −10 say. With subsequent growth there follows a rapid increase in complexity, and at birth the Number of Definition has increased to near 0 say. After birth there are probably several inflections in the curve, representing different phases in development (not shown on the graph), but with the coming of adulthood the Number of Definition finally tends to steady off at +10, say. The numbers given are purely illustrative.

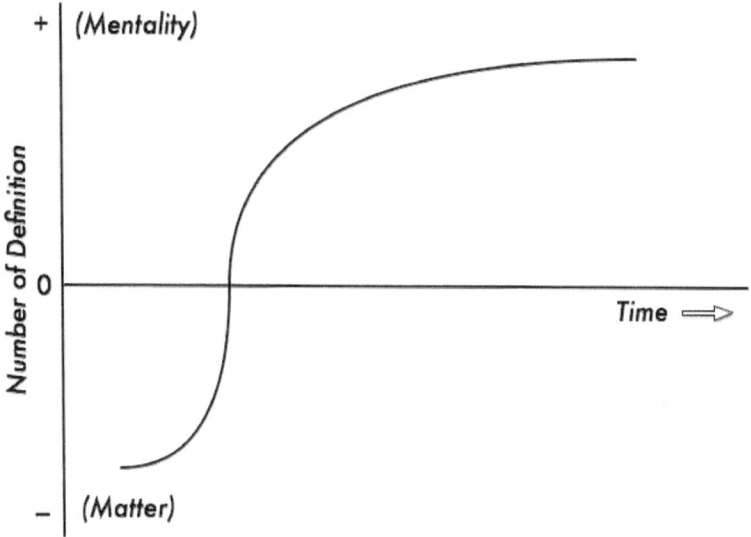

Figure 2.6. Evolution, Graph 3.

2.6.3. Coordination

The reader may have noticed that Energy was not directly represented in the previous diagrams. It is better to consider Energy on a separate diagram. However, we are more concerned with Energy Redistribution than with Total Energy itself, so we will only consider the Capacity for Energy Redistribution in the following discussion. There seems to be an optimum to this Capacity, depending on the Physical Size of the System considered. The following diagram illustrates this point. Figure 2.7 shows the relationship between the Capacity for Energy Control (ability to Redistribute Energy), and the Relative Size of the System.

MATTER, ENERGY AND MENTALITY

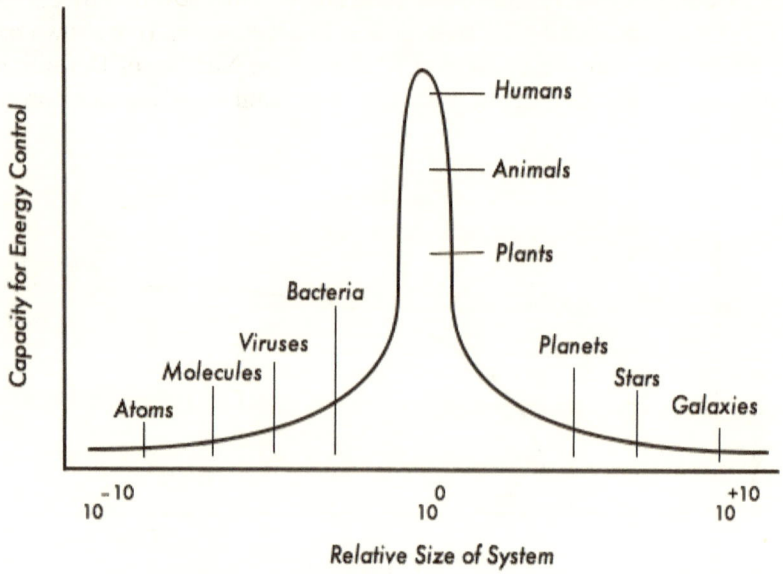

Figure 2.7. Capacity for energy control of various systems.

2.7. Conclusion

Living Systems arise more or less at random. With the phenomena of mutation and selection, Living Systems complexify. The more complex the system, the greater the capacity for Energy Control. When something is sought by the Organism, the means at its disposal require the intervention of Energy. The Organism is capable of selecting a certain pattern of events to the exclusion of other patterns of events. The flow of Energy that ensues is merely a Redistribution of the available Energy, resulting in a useful change for the Organism.

Chapter 3
Organism — The Source of Unnecessary Action

3.1. Living Systems

An Organism is the source of Unnecessary Action in its Environment. Any action not predicted by the Laws of Science but which, nevertheless, is subject to them, may be termed an unnecessary action. Such actions may, of course, be necessary for the Organism producing them. Action implies a flow of energy, and it is convenient to consider a meta-environment as the source of this energy. Thus an Organism diverts energy in the meta-environment to flow through its own environment. This is the essence of unnecessary action.

Practically every energy flow is accompanied by a total increase in entropy. Localised decreases in entropy must always be compensated for by greater increases in entropy in the meta-environment (Viswanadham, 1968). The question as to whether the entropy of the Universe as a whole, increases, decreases, or remains constant, is of no concern here. For details on this point see Popper (1965, 1967) and Büchel (1967).

Sir Julian Huxley (1963) writing about the Biosphere, has estimated that

> "The biological sector, considered spatially as the area occupied by life, cannot at the very outside constitute more than a million-million-millionth part of the extent of the visible Universe, and probably much less."

When discussing properties, which become measurable only in 1 part per 10^{18} of a system, their effects on the whole system can conveniently be ignored. However, turning a blind eye is not the answer, as this book attempts to show.

In this Chapter, localised decreases in entropy are considered to be brought about because of Energy Redistribution during energy flows. Now, since localised decreases in entropy are brought about during successive unnecessary actions, which by definition are unpredictable, they are independent of each other. That means that successive unnecessary actions are orthogonal. Thus, the Energy Redistribution brought about by an organism, during a suitable length of time may be considered as a coordinate of the source of action. In other words, it specifies the state of the Organism.

Table 3.1 summarises the Laws of Thermodynamics, including the less well-known Fourth Law.

Table 3.1. The Laws of Thermodynamics.

Zeroth Law
Bodies in contact with each other tend to reach thermal equilibrium.
First Law
Energy can be changed from one form to another but remains constant in a closed system or in the Universe as a whole.
Second Law
Entropy tends to increase in an isolated system or in the Universe as a whole.
Third Law
Entropy approaches zero as temperature approaches zero.
Fourth Law
Complexity tends to increase accompanied by localised decreases in entropy.

Living systems must be open to their environment to take in energy and material from outside. To remain compatible with the second law of thermodynamics, the overall entropy of the living system plus it's environment will increase during the process.

3.2. Treatment

Consider a space of n dimensions, with each coordinate being specified by the n various Energy Redistributions brought about during n successive unnecessary actions of equal duration T. Each coordinate can be represented by a vector in n-dimensional space, originating from the same point and ending on the circumference of a sphere, the radius of which is a measure of the average rate of Energy Redistribution produced by the Organism (and related to the Number of Definition considered in Chapter 2). Figure 3.1 illustrates this idea for a 3-dimentional space, where the diameter of the sphere represents the Maximum Capacity for Energy Redistribution of an Organism and the arrow represents one vector produced by one action of the Organism. The length of this arrow can be less than or equal to the diameter of the sphere.

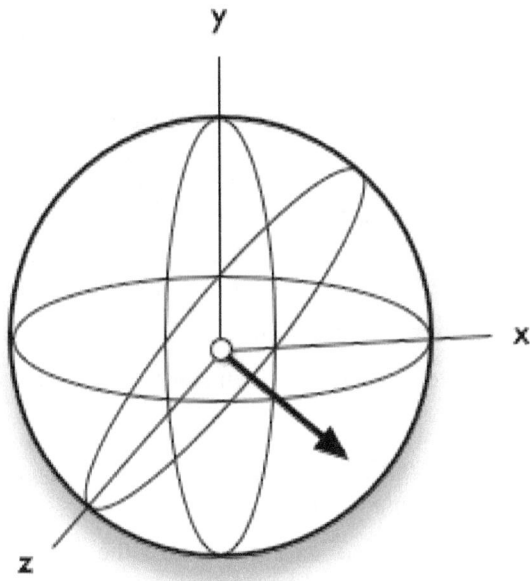

Figure 3.1. Representation of a 3-dimensional space.

Furthermore, if we quantize this space into concentric shells, we can assign probabilities to each shell. Such probabilities are the probabilities of the various vectors of duration T, which the source organism might produce to end somewhere on that shell surface. If the duration T is of sufficient length, practically all the vectors will end up somewhere on the same spherical shell. If, on the other hand, the time interval is reduced, then the distribution of the probabilities among the various concentric shells will increase. Some vectors will be longer, and some will be shorter, than the mean radius. It would appear, at first, that as the time interval is progressively reduced, the distribution of the vectors among the various shells should increase indefinitely – this is not so. There is an upper limit to the shell radius, which depends on the Organism, and is related to its Maximum Capacity for Energy Redistribution. No action, however short in duration, will produce a vector longer than this maximum value. In other words, there is no advantage in using durations less then those necessary to reach this limit. We can say that the actions of an Organism are uniquely and completely specified when this duration T_{min} is taken to specify each coordinate. Also there is no advantage in using a length of time greater than that which specifies the average or mean radius to a required degree of accuracy T_{max}. We can say that T_{max} specifies the central tendency, and T_{min} specifies the range of all the Unnecessary Actions produced by the Organism in question.

The above treatment follows the classical work of Hartley (1928) on the information content of continuous sources which revolutionised cybernetics (see also Shannon, 1949). However, I am more directly indebted to Cherry (1964) whose article enabled me to structure my thoughts in this section. In particular, I would like to bring to the attention of the reader, one of his key summarising statements,

> "Quantization is a logical necessity of specification."

He goes on to give an example which clarifies the point,

> "In this sense, symbols like π, e, $\sqrt{2}$, do not signify magnitudes at all, but rather have the nature of rules. Put another way, they do not have communicable magnitudes."

It follows that we are justified in quantizing the duration of action of an organism because the discreet coordinates thus formed, in themselves specify all that is required about the capacity for action of that organism. Furthermore, it is from an analysis of these actions that we can come to a conclusion about the Mentality or Consciousness of the Organism.

Chapter 4
Information and Entropy

4.1. Events and Observations

The field of information theory is an extension of thermodynamics and probability theory. Broadly, we can gain information in two distinct ways (Cherry, 1964):

Observation (Communication with Nature).

Communication in some agreed language or sign-system.

It follows that there are two main types of events:

Events by Observation.

Events by Definition.

The act of communication necessarily implies the existence of a set of probabilities. The information conveyed, transmitted, or stored, in a message is not an intrinsic property of the message itself. It depends on the set from which the message comes.

An event may be considered as a change in a definable variable. In any system, the event must not change the system irreversible, because such an event would, in a sense, be non-repeatable and transmute the variable. Thus, an event is a reversible or repeatable change in a definable system.

4.2. Information

Information is usually measured in binary digits called bits. The bit is the basic unit of information and one bit is the amount of information required to distinguish between any two equally possible alternatives. Mathematically, information is calculated as the negative logarithm to the base 2 of the probability of an event or outcome occurring.

Consider the amount of information I required to distinguish between two equally probable alternatives. By definition,

$$I = -\log_2(1/2)$$
$$= \log_2(2)$$
$$= 1 \text{ bit}$$

Hence, one bit of information is acquired when the value of a choice between two equiprobable possibilities becomes known.

More generally, if there are N equiprobable alternatives, the information required to specify any one of them is given as,

$$I = -\log_2(1/N)$$
$$= \log_2(N) \text{ bits}$$

When such a message is received, I bits of information are said to be gained by the recipient.

This can be generalized further. Any set of positive, not necessarily equal, fractions that add up to 1 may be regarded as a set of probabilities. For any set of probabilities, $p_1, p_2, p_3, \ldots p_n$, the following holds:

$$\sum_{i=1}^{n} p_i = 1$$

Shannon (1948) defined the Statistical Entropy S of such a set of probabilities as follows:

$$S = -p_1\log_2(p_1) - p_2\log_2(p_2) - p_3\log_2(p_3) \ldots -p_n\log_2(p_n)$$
$$= -\sum_{i=1}^{n} p_i \log_2(p_i)$$

Statistical entropy has two important properties:

> It has the maximum value when all the probabilities (n of them) are equal.

> Different individual values can be combined to yield an average entropy.

Such combinations are useful for finding the entropies associated with Markov chains. A Markov chain is a sequence of states in which, over various long periods, the probability of each transition is the same. The term is sometimes applied to a particular trajectory produced by a system, or the protocol of a particular matrix of transition probabilities. Table 4.1 illustrates an insect's Matrix of transition probabilities (between three states, Bank, Water, and Pebble) as described by Ashby:

Table 4.1. Matrix of transition probabilities.

↓	B	W	P
B	0.25	0.75	0.125
W	0.75	0	0.75
P	0	0.25	0.125

Each value in the Matrix is positive, in the range 0 to 1, and represents the probability of the transition indicated. All the values in each column also add up to 1. B = Bank. W = Water. P = Pebble.

Suppose that we start an experiment by placing 1000 insects under the Pebbles and then observe what happens. From a distance, we only see three possible populations, and the system as a whole is determinate, although the individual insects will behave with certain probabilities. It will be observed, over time, that the population tends to reach a state of equilibrium, ending up with around 449 insects on the Bank, 429 in the Water, and 122 under the Pebbles. These equilibrium values of Markov chains are readily computed as discussed in detail by W. Ross Ashby (1956). Another excellent and simpler introduction to Cybernetics is that given by Guilbaud (1961).

4.3. More About Information

In any system three types of information may be distinguished; structural information, I_s, functional information, I_f, and bound information I_b.

Structural information may be regarded as the information required to construct the system out of its defined component parts, or the information put into the system.

Functional information, on the other hand, may be regarded as the information required to perform the same function as the system in its environment-of-action. Consider the situation where the function of one system A is the construction of another system B. Then, depending on the relationship between I_f^A and I_s^B, three possibilities may be distinguished.

Case 1:
$$I_f^A < I_s^B$$
Most machines can be cited as examples of this group.

Case 2:
$$I_f^A = I_s^B$$
Apart from a number of interesting mechanical self-reproducing machines (Penrose, 1959), viruses are perhaps the best-known examples of this group.

Case 3:
$$I_f^A > I_s^B$$
Here, the whole range of living systems may be included, which not only reproduce themselves but also perform other functions as well.

A further type of information may also be specified. Wilson (1968a) makes the distinction between information bound into or hidden in a system I_b and information available to an observer outside the system,

> "An observer who knows the macrostate of a system has not yet obtained all the information he needs about the system. He can calculate its entropy, but he cannot specify its particular microstate."

Again, if a system has only one allowable microstate, the observer would know the state,

> "… not because he has maximum information about the system, but because he needs none."

This can be summarised by saying that bound information is required to specify the precise microstate of any resonant or flexible system.

4.4. More About Entropy

Because of a fundamental relationship between the definitions for information and entropy (Wilson 1968b), both structural and functional information, as defined above, may be measured in terms of entropy changes. In the former case, as the order put into the system (the entropy taken from the defined component parts, in the making of the system), and in the latter case, as the order put into (or maintained in) the environment-of-action by the operating system. In this way, positive values for both structural and functional information can usually be obtained, although this is not essential. For example, if the function of one system is the destruction of another system, it will have (using the above convention) negative functional information. On the other hand, bound information is always positive, and is best measured in terms of internal entropy units. For example, consider the rotation of carbon atoms with respect to one another in a carbon chain: the more branched the chain, the more hindered are the rotations. In other words, an increase in branching decreases the possible number of alternative (quantized) positions on the cones swept out by the various tetrahedral angles, resulting in a decrease in the internal entropy of the system (from the molecular point of view), and a corresponding decrease in the amount of bound information of the system (from the observer's point of view).

Woolhouse (1967) has cited an excellent example illustrating the difference between what we have called structural and functional information,

> "A single alteration in the relative positions of adjacent nucleotide bases could convert a crucial part of the genome code to nonsense so that its capacity to support development was lost, but this could occur without an increase, indeed even with a decrease, in the thermodynamic entropy content of the genome molecules."

Consider further the basic distinction between structural and functional information. Both are measurable in terms of entropy changes, but the former is independent of the environment, and is directly related only to the system itself and its defined parts, whereas the latter is independent of the system itself, but is directly related only to the effect the system is producing in its environment-of-action.

From the cybernetic point of view, the immediate environment may be considered as the specifier or selector of the precise conformational state of any flexible system. Consider such a system assuming various conformations in a changing environment. The structural information of the system will remain constant, provided that the changes are not too drastic of course. But, in some of these states, the system may manifest an ability to function. In other words, the environment may transduce the structural information of the system (when displayed in a particular conformation), into the functional information required to carry out a specific function in the environment-of-action. When this occurs the system works, and the information manifested by the system in carrying out its function is directly related to the entropy changes occurring (or being prevented from occurring) in the environment-of-action.

Viswanadham (1968) refuted Woolhouse's suggestion that living systems can be defined by the information content of their DNA-molecules alone:

> "Normal development depends on a congenial environment – environment being taken to include features such as temperature, humidity, radiation, air, water, food, and other living beings which can affect development."

Perhaps we could develop this, somewhat, by considering that living systems consist of a whole series of partially overlapping subsystems or microenvironments which interact with each other by transducing the structural information of portions of each other into the functional information required for the production and maintenance of some of the other portions ... and the organism as a whole.

This may be considered as a development of Woolhouse's discussion on the distinction and relationship between thermodynamic entropy and information content or the amount of developmentally meaningful organization in a system, as distinct from its structural organization.

Chapter 5
Three Lesser Laws of Reality

5.1. More about Energy Redistribution

In the preceding Chapters, Reality was analysed in terms of three fundamental entities—Matter, Energy and Mentality. It may be of interest here to consider some equations developed to describe some general features of Reality. However, before dealing with these equations in detail, the concept of the Differentiation of Energy needs to be reconsidered. Two forms of Energy were distinguished, namely the Physical and the Virtual forms. These are related as follows:

> Physical Energy determines the magnitude and Virtual Energy determines the direction of a more fundamental entity, and thus they have the ability to interact without disturbing the conservation of Physical Energy.

However, we will not be concerned with total Physical Energy, but only with the Capacity of Systems to Control this Energy. The Capacity of any system for Energy Control is defined as its ability to Redistribute Physical Energy by means of Virtual Energy, thus the amount of control depends on the amount of Virtual Energy operating during a Physical Energy flow.

It is further held that all Things or Systems in the Universe contain a certain finite amount of each of the three co-extensive entities of Reality—Matter, Energy and Mentality. And, it is not considered necessary that the total amounts of these three entities in the Universe is given by their algebraic sum in all the actual systems in existence at any given time. So, while it cannot be less than this, it may be more. But whatever these absolute amounts may be, they are considered constant.

From these basic ideas, it is possible to set up descriptive equations, which can be condensed into three fundamental ones, which are picturesquely called the Three Lesser Laws of Reality. These equations are not strictly mathematical but rather symbolic expressions, which, nevertheless, contain a certain amount of information, especially about the relationship between Matter, Energy and Mentality.

5.2. The First Lesser Law of Reality — Equation 1

The first equation is more of a definition than anything else, but a key one nevertheless. It relates Matter, Mentality and Energy Redistribution as follows:

$$M^*/M = E^\alpha$$

where M is the amount of Matter involved in the vicinity of the system, E^α is the Energy Redistribution brought about by the system and M^* is the Mentality required to bring about this redistribution.

Taking logarithms on both sides (not to any particular base) gives the following result:

$$\log(M^*/M) = \log(E^\alpha) = \text{Number of Definition}$$

The First Lesser Law of Reality may then be stated as follows:

> The Number of Definition of any System is the logarithm of the ratio of Mentality to Matter involved, and this is related to the logarithm of the Energy Redistribution.

This Number of Definition of a System is not precisely defined by a single action but, rather, by the overall Energy Redistribution brought about by the System over a suitably appropriate length of time.

This leads on to a second equation.

5.3. The Second Lesser Law of Reality — Equation 2

Every system has a Maximum Capacity for Energy Redistribution E^α_{max}, which depends on its Relative Size and its Organizational Complexity.

5.3.1. Maximum Capacity and Relative Size

As systems become very large or very small, the Capacity for Energy Redistribution decreases (as illustrated in Figure 2.7 in Chapter 2). The minimum value approaches zero in both directions, but it never becomes negative. For this reason, only positive values of $\log(m_{rel})$ are considered, and negative numbers are converted to positive numbers — these are called absolute values and are indicated by two vertical lines.

So, the Maximum Capacity for Energy Redistribution can be hypothetically related to the Relative Size of a System as follows:

$$E^\alpha_{max} = a / |\log(m_{rel})|$$

where the symbol α is used to represent Energy Redistribution, a is a constant, and m_{rel} is the Relative Size of the System, taking the Human as an appropriate unit size, and $|\log(m_{rel})| \geq 0$. The value of the above expression approaches infinity as $|\log(m_{rel})|$ approaches zero, and that fits in with our general discussion; however, the value at exactly zero is mathematically indeterminate.

At the same time, the relationship between the Actual Energy Redistribution of a System and its Maximum Capacity for Energy Redistribution is given as

$$0 \leq E^{\alpha}_{act} \leq E^{\alpha}_{max}$$

So, it follows, that

$$E^{\alpha}_{act} \leq a / |\log(m_{rel})|$$

Thus far, The Second Lesser Law of Reality may be stated as follows:

> The Capacity for Energy Redistribution of any System is inversely related to the absolute value of the logarithm of the Relative Size of the System.

5.3.2. Maximum Capacity and Organizational Complexity

The Maximum Capacity for Energy Redistribution of any System depends not only on its Relative Size but also on its Organizational Complexity.

Again, the Maximum Capacity for Energy Redistribution can be hypothetically related to the Organizational Complexity of a System as follows:

$$E^{\alpha}_{max} = b \log(W)$$

where b is a constant and $\log(W)$ is the Organizational Complexity of the System, and $\log(W) \geq 0$.

This part of The Second Lesser Law of Reality may be stated as follows:

> The Capacity for Energy Redistribution of any System is directly related to the Organizational Complexity of the System.

It also follows that

$$E^{\alpha}_{act} \leq b \log(W)$$

MATTER, ENERGY AND MENTALITY

Combining both hypothetical relationships of The Second Lesser Law of Reality gives the following result:

$$E^{\alpha}_{max} = a/|\log(m_{rel})| = b\log(W)$$

Therefore

$$a/b = \log(W)\,|\log(m_{rel})|$$

Letting $a/b = c$ another constant, gives the result:

$$c = \log(W)\,|\log(m_{rel})|$$

This may be stated as follows:

> The Organisational Complexity and the absolute value of the logarithm of the Relative Size of a System vary inversely in the ranges permitted.

In general, both parts of The Second Lesser Law of Reality may be combined into the following statement:

> The Capacity for Energy Redistribution of any System is inversely related to the absolute value of the logarithm of the Relative Size of the System, and is directly related to the Organizational Complexity of the System.

The term $\log(W)$ is also used in Thermodynamics where Statistical Entropy S is related to it as follows:

$$S = k\log(W)$$

where k is another constant. If natural logarithms are used, the constant becomes Boltzmann's constant k_B.

$$S = k_B \log_e(W)$$

Entropy was considered previously in Chapter 3.

Now, compare the following equation:

$$I = k\log_b(N)$$

This is a formula for the amount of information I conveyed in a message selected from a set of possibilities N, where the integer $N \geq 1$ and the constant $k > 0$ and depends on the value of the logarithmic base b. The similarities between Statistical Entropy and Information thus become apparent.

If $b = 2$ then $k = 1$, and the equation reduces simply to,

$$I = \log_2(N)$$

Furthermore, if N = 1 then $I = 0$, because there is no additional information required to specify the state of the system in this trivial case.

5.4. The Third Lesser Law of Reality — Equation 3

If the ratio of the Total Amounts of Mentality and Matter in the Universe at any time t is represented by ω_t, then,

$$\frac{M^*_{total}}{M_{total}} = \omega_t$$

Taking logarithms on both sides gives the following:

$$\log\left(\frac{M^*_{total}}{M_{total}}\right) = \log(\omega_t)$$

In the limit, as t gets very large,

$$\lim_{t \to \infty} (\omega_t) = \Omega$$

Another way of writing the above equation is as follows:

$$\lim_{t \to \infty}\left[\sum_{i=1}^{N_t}\left(\frac{M^*_i}{M_i}\right)\right] = \Omega$$

where N_t is the number of Systems in existence in the Universe at time t, and the other terms have their usual meanings.

Similarly,

$$\lim_{t \to \infty}[\log(\omega_t)] = \log(\Omega)$$

or more fully,

$$\lim_{t \to \infty}\left[\sum_{i=1}^{N_t}\log\left(\frac{M^*_i}{M_i}\right)\right] = \log(\Omega)$$

where $\log(\Omega)$ is the Ultimate Number of Definition of the Universe.

Table 5.1 summarises three possible situations, depending on how much Mentality and Matter is involved in the Universe.

Table 5.1. Summary of three main possibilities.

Universal Ratio	Ω	Log (Ω)
$M^*_{Universe} < M_{Universe}$	< 1	Negative
$M^*_{Universe} = M_{Universe}$	1	Zero
$M^*_{Universe} > M_{Universe}$	> 1	Positive

Overall, the Third Lesser Law of Reality may be elaborated as follows:

> An extension of the First Lesser Law of Reality can be applied to the entire Universe.
>
> If the ratio of the Total Amounts of Mentality to Matter engaged in the Universe at any time t is ω_t, then the Number of Definition of all the Systems in the Universe at that time is log (ω_t).
>
> The number of Systems in the Universe tends to increase with time.
>
> The Organisational Complexity of many Systems in the Universe also tends to increase with time.

All this may be summarised into the following statement of The Third Lesser Law of Reality:

> The Number of Definition of the Universe is the logarithm of the ratio of Mentality to Matter involved in the Universe, and this is related to the logarithm of the Maximum Capacity for Energy Redistribution of all the Systems in the Universe, and the Number of Definition of the Universe tends to increase with time until an Ultimate Value is reached.

The best way of looking at this Third Lesser Law of Reality is to consider it as an expression of the tendency of things to evolve in the Universe. We are accustomed to thinking of evolution solely as a biological phenomenon; however, it is much more (Teilhard de Chardin, 1966; Hoyle, 1963; Huxley, 1963). In a comprehensive sense, it includes all aspects of the Universe, from the highest biological level to the lowest inorganic level and, in this book, it even includes a rather novel aspect which seems to lie outside these limits; I refer to the evolution of the Universe itself from the three basic Sectors of Reality, Matter,

Energy and Mentality.

We are also accustomed to thinking that the Universe contains a definite amount of Matter, from which everything else ultimately arises, including Mentality, when a system is elaborate enough. However, there are other ways of looking at the Universe. The Universe itself may arise, at least partially at first, from the three fundamental overlapping Sectors of Reality, and from there Systems arise and complexify, revealing the overlap.

5.5. Summary

In the treatment of the three Lesser Laws of Reality, given above, several points emerge. These points can be summarised non-mathematically as follows:

> It was found necessary to construct a working definition of Mentality suitable for analysis. This definition is called the First Lesser Law of Reality and it converts the problem of defining Mentality into a problem of defining Energy Redistribution.
>
> Two types of Energy were identified (in Chapter 2), Physical Energy and Virtual Energy; the former having its usual meaning and the latter being the agent for the redistribution of the former, resulting in unnecessary action under the direction of Mentality.
>
> The Second Lesser Law of Reality relates the capacity for Energy Redistribution to the Relative Size of a System. The Human was proposed as the unit size since the inverse relationship of the Second Lesser Law of Reality approaches an infinite Capacity for Energy Redistribution for an organism of unit size. Objects larger or smaller than this have a progressively reduced capacity for Energy Redistribution.
>
> The Second Lesser Law of Reality also relates the Capacity for Energy Redistribution to the degree of Organisational Complexity of a System.
>
> An expression relating the Relative Size and Complexity of a System to the Capacity for Energy Redistribution was derived.
>
> The Third Lesser Law of Reality anticipates that in the course of time, Systems should arise and evolve to an ultimate steady state of equilibrium in the Universe. This equilibrium state is reached when the Number of Definition of the Universe reaches $\log(\Omega)$, the value of which depends on the absolute total amounts of Mentality and Matter available for the ultimate construction of the Universe.

5.6. Conclusion

Three Lesser Laws of Reality were formulated from the above considerations as follows:

The First Lesser Law of Reality

The Number of Definition of any System is the logarithm of the ratio of Mentality to Matter involved, and this is related to the logarithm of the Capacity for Energy Redistribution.

The Second Lesser Law of Reality

The Capacity for Energy Redistribution of any System is inversely related to the absolute value of the logarithm of the Relative Size of the System, and is directly related to the Organizational Complexity of the System.

The Third Lesser Law of Reality

The Number of Definition of the Universe is the logarithm of the ratio of Mentality to Matter involved in the Universe, and this is related to the logarithm of the Maximum Capacity for Energy Redistribution of all the Systems in the Universe, and the Number of Definition of the Universe tends to increase with time until an Ultimate Value is reached.

Appendix 1
Logarithms, Summations and Limits

A.1.1. Logarithms
The following logarithmic relationships apply if all the bases remain the same.
1. Change a multiplication into an addition:
$$\log_b(mn) = \log_b(m) + \log_b(n)$$
2. Change a division into a subtraction:
$$\log_b(m/n) = \log_b(m) - \log_b(n)$$
3. Change an exponent into a multiplier:
$$\log_b(m^n) = n\log_b(m)$$
4. Change a root into a divider:
$$\log_b(\sqrt[n]{m}) = \log_b(m^{-n}) = [\log_b(m)]/n$$

The following relationship can be used to change the base of a logarithm.

5. Change a base into another base:
$$\log_b(x) = \log_d(x)/\log_d(b)$$

The term $1/\log_d(b)$ is a constant, so the above equation may be written as follows:
$$\log_b(x) = k \log_d(x)$$
where k is a constant.

The three most commonly used logarithm bases are 10, e (approximately 2.718) and 2.

Logarithms to the base 10 are called common logarithms: $\log_{10}(x)$. Units of decimal logarithms are called decimal units or digits (sometimes dits, Hartleys or bans).

Logarithms to the base e are called natural logarithms: $\log_e(x)$ or ln (x). Units of natural logarithms are called natural units or nats (sometimes nepits or nits).

Logarithms to the base 2 are called binary logarithms: $\log_2(x)$. Units of binary logarithms are called binary digits or bits.

MATTER, ENERGY AND MENTALITY

A.1.2. Summations

The symbol for a summation is the Greek capital letter sigma, Σ.
For example, the summation of a series of n terms, $p_1, p_2, p_3 \ldots p_n$, may be written simply as follows:

$$\sum_{i=1}^{n} p_i$$

A.1.3. Limits

The abbreviation for a limit is lim.
For example, the limit of the series, $x_1, x_2, x_3, \ldots x_n$, as n approaches infinity, may be written simply as follows:

$$\lim_{n \to \infty} (x_i)$$

Appendix 2
Information and Observation

A.2.1. Observation of a System that Always Changes State

Consider a system that can assume N different possible states. It will have $(N-1)$ ways of changing from one state into another or of discontinuing in its current state. So, each change of state or event will require $\log_2(N-1)$ bits of information to be specified, because the initial state is known and excluded from being repeated in a change of state. However, the initial state could be any of N possibilities; so, the information required to know the initial state is $\log_2(N)$. Therefore, the total information required to specify a change involves two states and is given as follows:

$$I_1 = \log_2(N) + \log_2(N-1)$$

Table A.2.1 summarises the typical characteristics of this type of system.

Table A.2.1.

Information involved in a system that always changes state.

Number of different states	N
Number of ways of changing state	$N-1$
Information gained per observation of a change or required to specify any one change	I_1 $= \log_2(N) + \log_2(N-1)$ $= \log_2[N(N-1)]$ $= \log_2(N^2 - N)$
Total number of changes observed	T
Information required to specify any T changes or gained from observing such changes	$I_{T(Random)}$ $= T[\log_2(N) + \log_2(N-1)]$ $= T\log_2(N) + T\log_2(N-1)$ $= \log_2(N)^T + \log_2(N-1)^T$
Information required to specify T sequentially ordered changes or gained from observing such changes	$I_{T(Ordered)}$ $= \log_2(N) + (T-1)\log_2(N-1)$ $= \log_2(N) + \log_2(N-1)^{(T-1)}$
Information required to specify all T sequentially ordered changes that could possibly occur or be observed	$I_{\sum T(Ordered)}$ $= T![\log_2(N) + (T-1)\log_2(N-1)]$ $= \log_2(N)^{T!} + \log_2(N-1)^{(T-1)T!}$

A.2.2. Observation of a System that Sometimes Changes State

Observation of a system at intervals over time allows the possibility of the same state occurring more than once in succession. It is, therefore, an expansion of the idea of successive change to include the idea of successive continuity.

Consider a system that can assume N different possible states. It will have 1 way of continuing in its current state and $(N - 1)$ ways of changing to another state. So, the total number of possibilities per observation will be given as:

$$1 + (N - 1) = N$$

and each observation will gain $\log_2(N)$ bits of information.

Table A.2.2 summarises the typical characteristics of this type of system.

Table A.2.2.
Information involved in a system that sometimes changes state.

Number of different states	N
Number of possible states per observation	N
Information gained per observation or required to specify any one state	I_1 $= \log_2(N)$
Total number of observations	T
Information required to specify any T states or gained from observing such states	$I_{T(Random)}$ $= T\log_2(N)$ $= \log_2(N)^T$
Information required to specify any T sequentially ordered states or gained from observing such states	$I_{T(Ordered)}$ $= T\log_2(N)$ $= \log_2(N)^T$
Information required to specify all T sequentially ordered observations that could possibly occur	$I_{\sum T(Ordered)}$ $= T!\log_2(N)^T$ $= \log_2(N)^{T\,T!}$

Appendix 3
More about Functional Information

A.3.1. Biological Functional Information Distributions

At this point, it may be of interest to suggest a case where even whole groups of cells in a tissue can exhibit a variable efficiency, under apparently identical experimental conditions. From a series of 147 experiments by Davenport (1952) on gastric acid secretion (each measuring simultaneous secretion rates of hydrochloric acid (HCl) and oxygen (O_2) consumption by the oxyntic cells of the frog gastric mucosa), it was strikingly shown that a frequency distribution of the values of the ratio ⊔ HCl/⊔ O_2 occurred (Conway, 1959). Other tissues also show this type of distribution (Davenport & Chavré, 1953). Whatever the explanation for these controversial observations, it is unlikely that they can be explained in terms of experimental error alone. They are cited here not as proofs, but as possible illustrative examples of what appear to be biological functional information distributions.

A.3.2. Active Transport: A Biophysical Example

We are now in a position to apply the concepts of information elaborated in Chapter 4 to various systems. Let us consider active transport as a working example. Again, for simplicity, and following Spanner (1954), let us confine the discussion to a completely isolated system containing water and a number of solutes divided into two equal halves by a semipermeable membrane. Imagine the system, at first, to be in absolute equilibrium, so that all of its constituents are uniformly distributed everywhere.

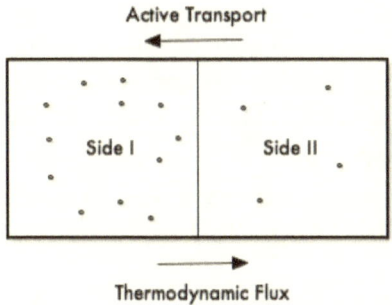

Figure A.3.1. An isolated system containing solutes distributed across a semipermeable membrane.

A disturbance in the equilibrium may be caused by an activity of the membrane, which causes a redistribution of one, or more of say n solutes within the system. If we denote the final distribution (difference in concentration across the membrane) of the *i*th species by Δx_i, this may be specified as follows:

$$\Delta x_i = (x_i^I - x_i^{II})$$

Then, provided no thermal, electrical, chemical or other disturbances have occurred in the system, the final state of the system can be fully specified by the variables, $\Delta x_1, \Delta x_2, \Delta x_3, \ldots \Delta x_n$. Such a system will tend to increase its entropy, and the rate of increase may be expressed as follows:

$$\frac{dS}{dt} = \sum_{x=1}^{n} (\Delta x_i j_i)$$

where *n* is the number of different solutes and each j_i is the force tending to change each Δx_i back towards equilibrium at zero.

Now, in order to maintain this disturbed state, the membrane must either become completely impermeable, or continue to redistribute the solutes as they return across the membrane towards equilibrium. In the former case, two independent closed systems would be set up, so that solution is inadmissible, or at best a trivial case.

However, if the membrane remains semipermeable, it must continue to supply functional information to the environment. When this type of system reaches a steady state, the rate of flow of functional information into the environment, exactly balances what the rate of increase of entropy would otherwise be in that same environment.

This functional information-supplying process is known to reside in a number of specific enzyme-carrier complexes in the membrane. Let us say, N_1 of them are responsible for transport of the first species, N_2 of them for the second species, ... and N_n of them for the nth species.

The average rate of transmission of functional information to the environment-of-action per carrier, averaged over all types of carriers in such a system, will then be given by the following equation:

$$d(\overline{I_f})/dt = \sum_{i=1}^{n}(x_i j_i) / \sum_{i=1}^{n}(N_i)$$

where I_f (bar) represents the average functional information supply per carrier and the other symbols have their usual meanings.

Viswanadham (1968) has criticised Campbell (1967) for basing his 'Biological Entropy Pump' on the possibility of a statistical distribution of entropy among a large number of entities – against which theoretical objections can be raised. Now while this criticism is valid, as far as it goes; nevertheless, the basic concept underlying Campbell's theory should not be dismissed.

To elaborate (again using active transport as an example), any cell could survive with only a fraction of its normal quota of carriers, provided that this fraction would operate at its maximum efficiency all the time. However, since this is obviously not possible in practice, an excess of carriers is produced, and these can operate at differing rates (provided a minimum average rate is satisfied). Consider the simplest case where only one species is being transported on one set of carriers: here all the x_i's are the same, but not necessarily all the j_i's. This would give rise to an unequal distribution of functional information, among all the n_i carriers. Now imagine what would happen if such a cell was subdivided into a number of smaller ones, each containing only a fraction of the carriers from the parent population. Under these conditions, not all the cells formed would have enough functional information to survive independently, whereas others (with the same number of carriers) would. This follows, because the statistical distribution of functional information among a large number of entities may lead to a small percentage of them possessing a higher information content than the mean value of the group as a whole, even though the structural information (the thermodynamic entropy if you like) of each is the same. And this applies to carriers, codfish eggs, or indeed any flexible functional system.

Appendix 4
Different Levels of Organization

A.4.1. Organisation

In this Appendix, various information-entropy and information-variety interfaces are considered. Three broad, but partly overlapping, categories of organization are distinguished; namely,

Abstract Organisation.

Physical Organisation.

Biological Organisation.

A.4.2. Abstract Organization

Any set of possible values z_i where $i = 1, 2, \ldots l$, of a variable z, has a variety or associated nescience given as follows:

$$(V_z)_j = -\sum_{i=1}^{i=l} (p_i)_j \ln (p_i)_j$$

The various possibilities z_i, can be represented by any mathematical set of fractions provided $0 \leq (p_i)_j \leq 1$; and $\Sigma (p_i)_j = 1$. The set may, as in human communication, represent the probability of receiving various answers to some well-defined question. If all the possible answers are not known *a priori*, then according to Tribus & McIrvine (1971), one does not have a properly defined question. Sometimes all the answers are known, but nothing more. In such an initial state of nescience one would have to assign equal probabilities of $(p_i)_0 = 1/l$ to all the answers. Now let some information be transferred such that the new probabilities $(p_i)_1$ can be assigned to the various members of the set, enabling a better inference to be made about the actual value of z. The information carried by this first step is given as the difference $(V_z)_0 - (V_z)_1$. Proceeding in this way, eventually after j steps say, the single correct answer will be known, where all but one of the $(p_i)_j$ values are now zero. The nescience about the topic or system has therefore been reduced from $\ln (l)$ to $\ln (1) = 0$, which represents a total information of l nat gained about the variable z. (Here, nat are natural units of information using logarithms to the base e, and ln is short for $\log_e$). Each of the j steps involved in gaining this information represent progressively more organized stages of knowledge about the system.

It should be noted that physical entropy does not play a part in such systems. Admittedly, energy and entropy fluxes are involved during the search for the correct answer, but such involvement is peripheral; it is a consequence of the method used in finding the answer, not the information carried by it. In abstract organization, the structure or logon content—a term coined by MacKay (1950), which may be defined as $V_{z\,max} = \ln l$—does not change once it is formulated. Only the probabilities assigned to the different answers during the search can vary. The concept of physical entropy, in these cases, is replaced by an analogous concept of associated variety or nescience.

In general,

$$A_z = Q_z / V_{z\,max}$$

where $V_{z\,max} = \ln l$ and l is the number of possibilities in question. This is an important result proved by Khinchin (1957). But $Q_{z\,min} = 0$ and $Q_{z\,max} = \ln l$. Hence $0 < A_z < 1$ for all abstract systems. Indeed, as A_z changes from the initial value 0 to the final value 1, the divergence of the p_i values from equiprobability increases. Gatlin (1972) has denoted such divergence as D_1-type divergence.

A.4.3. Physical Organization: Machines

According to Büchel (1967),

> "Whenever a mechanism containing an information of n bits is built, the thermodynamic entropy of that mechanism or its environment must increase by the amount of at least k n log 2, where k is Boltzmann's constant."

Büchel was considering a watchmaker putting the separate parts of a watch together to make a functioning mechanism. He was wrong in suggesting that Boltzmann statistics may sometimes hold for such a small number of many different, somewhat arbitrary shaped, and relatively enormous chunks of matter! Here again, the reduction in variety of the defined parts during construction has little or nothing to do with physical entropy. Different watches require different amounts of dissipated energy to attend their construction. However, even with the most conservative designs, the parts are not remotely like molecules. A single cog is almost as abstract an item as the numbers indicating the directions of the hands of the completed watch. Both have almost no functional existence outside a watchmaker's (or a watch user's) mind! The main difference between a box of loose parts and the finished working model is that the latter can store and release energy in a regulated way. But, once again, there is no relation between the physical energy and entropy fluxes and the information gained by a timekeeper. In any case, the variety of the separate parts (as a group) is much greater than that of the assembled watch and if the entropy argument can be applied at all, the entropy of the watch must decrease on construction. Admittedly, the entropy of the environment will increase by at least the same amount.

A more important point is the fact that when the watch is completed it becomes a completely determinate machine. To the nearest second, the machine generates 43,200 distinct states over a period of 12 h. An alien information theorist observing these states would gain information about the machine during the first 12 h, and thereafter would probably conclude that the machine was determinate, and therefore could not generate any new information.

This situation is analogous to the bookbinder's error described by Cherry (1957), where every page of a book turned out to be identical. The behaviour of the watch can be fully described by a matrix of 43,200 rows and columns where all transitions have a probability of occurrence of 0, except on the diagonal where all the values are 1. The result is a closed first-order Markov chain. If all transitions were equiprobable (that is independent of the previous states of the system), they would each occur with a probability of 1/43,200. The trajectory of such a process would describe an open zero-order Markov chain. Gatlin (1972) has termed divergence from independence as D_2-type divergence.

Penrose (1959) described some interesting self-reproducing machines which assembled parts, when agitated, according to their own likeness. Such toys certainly reduce the variety of the parts around them, but the same argument as in the above example applies.

Higher order Markov sources are well known. For example, the letters of the written English language (as used in this paper) may be shown to contain inter-letter correlations reaching back to cover at least eight letters preceding them (Hassenstein, 1971).

A.4.4. Biological Organization: Cells

According to Paul (1964), there are three separate aspects of organization in cells namely:

>Organization for propagating the genetic information.

>Organization for synthesis of macromolecules.

>Organization whereby energy is constantly supplied to maintain steady state or homeostatic conditions.

Each aspect will now be briefly considered.

A.4.4.1. Genetic Information

The structural organization of genetic information is ultimately contained in DNA. But according to Warburton (1967),

> "The genetic DNA of a highly evolved organism carried no more information than a random concatenation of the same number of nucleotide pairs."

It is only in its functional organization that DNA becomes selective in determining what protein systems operate in the cell. This is clear when one compares a DNA molecule with its corresponding free nucleotides; the genetic information carried by the free nucleotides is nil. Thus, the variety of the nucleotide sequence only represents one type of organization—structural organization; whereas, the variety of cellular expression represents a different type of organization—functional organization. In such cases, the structural and functional organization derive from different systems, and can have no mathematical relation to each other.

Mutations, which occur during replication, may be produced because of the equivalent of noise at the molecular level of the genetic landscape. The presence of high concentrations of such simple cations as Mg^{++} and NH_4^+ can cause misreading of various codon assignments. It can even be argued that a certain amount of background noise may be of evolutionary advantage. However, at a purely molecular level replication can be considered in the same way as any other macromolecular synthesizing system.

Gatlin (1968) has applied the distinction between D_1- and D_2-type divergence to the study of DNA sequences. She considered the four bases: adenine A, cytosine C, guanine G, and thymine T, to represent an alphabet, and their sequence on a DNA molecule to represent a first-order Markov chain. One important distinction to emerge between vertebrates and lower organisms was that vertebrates tended to hold D_1 constant and increase D_2, whereas lower organisms tended to do the opposite.

Each base pair can be one of four possible types, with Adenine always pairing with Thymine and vice versa, and Guanine always pairing with Cytosine and vice versa. So, each base pair encodes exactly $\log_2(4) = 2$ bits of information. In the human body, the haploid genome in sperm or egg cells contains 3.2 billion DNA base pairs, and the diploid genome in somatic cells contains 6.4 billion pairs. Therefore, the haploid human genome contains about 6.4 gigabits of information and the diploid genome contains 12.8 gigabits.

A.4.4.2. Macromolecules

Ryan (1975) outlined a method for classifying macromolecules (including DNA) according to their structural, functional and bound information values. The two terms structural and bound information may replaced by the term transitive information, which has selective, connective and conformative components. These components represent different stages of structural organization (Ryan, 1975). Functional information, on the other hand, may be regarded as the information required to perform the same function as the macromolecule or any other system in its environment-of-action.

A.4.4.3. Steady State Systems

Steady state or homeostatic systems, are perhaps the most characteristic of biological systems, and are never at equilibrium. According to Angrist & Hepler (1973),

> "Life, the temporary reversal of a universal trend towards maximum disorder, was brought about by the production of information mechanisms."

A good example of such a mechanism is active transport across cell membranes. According to Holden (1968), transport systems were perhaps the first (extragenetic) informational systems to evolve. They may have even preceded enzymes in this regard. The quantitative aspects of simple diffusion were first worked out by Fick (1855). A large number of transport systems, however, do not follow Fick's equation. They can only be given satisfactory treatment by an extended diffusion theory, which considers the effect of other forces. The theory was originally formulated by Onsager (1931), who subsequently received the Nobel Prize in 1968 for this contribution. Essentially, this provided the first satisfactory treatment of non-equilibrium thermodynamic systems.

According to Ptitsyn & Birshtein (1969),

> "It is very probable that biological molecules or their complexes reversibly (cyclically) pass from one conformational state into another in the process of their functioning."

From this, it can be concluded that a functioning system is not in equilibrium. If the point of equilibrium is too far removed, then the above treatment cannot be applied because the fluxes J_i, are then too fast to remain linearly proportional to their conjugated forces X_i.

MATTER, ENERGY AND MENTALITY

Fortunately, it is possible to derive an expression for the variety (information) required to maintain such a process without making any assumptions as to the physical nature of the system (whether Boltzmann-like or otherwise). This may be achieved by constructing a Markov matrix of transition probabilities at the steady state.

A.4.4.4. A Worked Example: Active Transport Again

As an illustration, consider the following simple case:

Table A.4.1. Matrix of transition probabilities.

↓	A	B	C
A	0.87	0.09	0.84
B	0.13	0.60	0.11
C	0	0.31	0.05

Each value in the Matrix represents the probability of the transition indicated. All the values in each column add up to 1. A = On Side A. B = On Side B. C = On the Carrier.

Here, only one solute is considered which may be in any of three locations: side A, side B, or attached to a carrier C. From inspection of the matrix, it may be seen that a solute molecule on side A tends to diffuse to side B (depending on such factors as the electrochemical gradient and the relative permeability of the membrane). It cannot, however, attach to a carrier from side A. If the solute molecule is on side B, it has the greatest tendency to stay there, but also has a chance of combining with a carrier (depending on their number and the collision frequency). It has the least chance of passing directly back across the membrane (by say simple diffusion against a concentration gradient). The transition probabilities may include many factors such as particle-particle interactions, if present in the system.

An experiment can be started by adding an amount of solute molecules, say 200 mmol, to either compartment A or B, and waiting until conditions have settled down to a steady state. Ashby (1965), and many others, have shown how an equilibrium state can be derived from a transition matrix. (The converse procedure is of course also possible). Basically, this is achieved by solving three simultaneous equations which, in the present example, are as follows:

$$N_A = 0.87N_A + 0.09N_B + 0.84N_C$$

$$N_B = 0.13N_A + 0.60N_B + 0.11N_C$$

$$N_C = 0.00N_A + 0.31N_B + 0.05N_C$$

where N_A, N_B and N_C, are the amounts of solute in the three locations at equilibrium. Volume considerations have been omitted for simplicity. Also,

$$N_A + N_B + N_C = 200 \text{ (mmol)}$$

at all times. Therefore, the required values at equilibrium can be found to be (approximately): $N_A = 136$, $N_B = 48$, and $N_C = 16$ mmol, respectively.

These values, including the net binding of solute molecules to the carriers (or at least to the cell wall/plasma membrane), can be determined experimentally. (Some of these aspects were considered by Ryan, *et al.*, 1971, and Ryan & Ryan, 1972).

Furthermore, the associated variety of the system may be calculated from the defined space of 2-tuples: $p_i p_j$ where i, j = A, B, C. For the above first order Markov process then, the associated variety $V_{M(1)}$, is given by the sum of nine terms as follows:

$$V_{M(1)} = - \sum_{i=1}^{i=C} \sum_{j=1}^{j=C} p_i p_{ij} \log_e \left(p_i p_{ij} \right)$$

where $p_A = 135/200 = 0.68$, $p_B = 49/200 = 0.24$ and $p_C = 16/200 = 0.08$. The nine p_{ij} values are taken from the matrix, and the final result is 1.14 nat. The ploy used is analogous to that for non-parametric statistical methods when no *a priori* assumptions are made about the possible data distribution.

The method is readily extended to enzyme systems. Quastler (1972) discusses some of these aspects. Now the maximum possible variety associated with such a system is given when all p_i and p_j values become equiprobable (that is equal to 1/3). The maximum variety, therefore, is ln (9) = 2.20 nat. Finally, The difference 2.20–1.14, or 1.06 nat, represents the actual functional information required per particle of the system. There are 0.2 x 6.02 x 10^{23}, or 1.20 x 10^{23} such particles, requiring a total of 1.28 x 10^{23} nat per unit time for the whole system: this being generated by the active transport carriers. As suggested, this value can be expressed per carrier, and is the non-parametric information equivalent of the physical entropy change involved. Thereby, under certain circumstances, it is possible that exact values for a proportionality constant k_x may be determined for different systems.

A.4.5. And Finally ...

Finally, biological systems, because of their high levels of organization, exhibit two unique effects:

Cellular Information Processing.

Ability to Codify Information.

A.4.5.1. Cellular Information Processing

The unicellular bacterial organism *E. coli* is capable of processing up to 2 knat/s of information (value derived from data of Lwoff, 1965). Taking this value as the least value representative of a living cell, consider a hypothetical organism consisting of say 50 such cells. Part of its 100 knat/s information processing capacity is required by the organism to keep itself together. For example, if each cell can occupy any of 50 possible positions, relative to any other cell, then ln (50!), or just 148 nat (say, per second also, for convenience) of information is required for the process. If some of the cells are differentiated, this value will become much lower because not all of the positions will then be allowable. Relative orientations of each cell have been ignored because undifferentiated cells do not normally require a 'this-side-up' specification. Therefore, 99.85 knat/s will still be available for other processes. Thus, even for such a small organism, the relative amount of information required to specify microstructure is almost negligible. This example is over simplistic, of course—organisms are not normally made by picking and placing separate cells. In reality, differentiation during development dictates location and *vice versa*.

A.4.5.2. Ability to Codify Information

Miller (1958) refers to a second effect: the ability to codify information at higher levels of organization. Briefly, this means that certain channels (or their terminals) have the ability to convert several 'bits' of information into single 'chunks', which are not amenable to classical Shannon treatment. This effect is the first stage in formulating the components of abstraction. Such components or chunks have progressively less direct physical relation to their entropic sources. In this context, the 'bit/chunk' ratio may serve as a good index of the system's departure from physical entropy. It is concluded that rigorous definition of information-entropy interfaces at different levels of organization will lead to a useful classification of diverse systems within a single working framework.

Bibliography

Angrist, S.W. & Hepler, L.G., Order and Chaos: Laws of Energy and Entropy, Harmondsworth: Penguin Books Ltd. (1973).

Ashby, W.R., An Introduction to Cybernetics, Chapman & Hall Ltd., London, (1956). pp. 165–169.

Avery, J.S., "Chapter 4: Statistical Mechanics and Information", In: Information Theory and Evolution, World Scientific Pub. Co. Pte. Ltd. (2003). pp. 73–94.

Barrow, J.D. & Tipler, F.J., The Anthropic Cosmological Principle, (Forward by Wheeler, J.A.). Oxford: Oxford University Press. (1986).

Ben-Naim, A., "1.3. The Association of Entropy with Missing Information", A Farewell to Entropy: Statistical Thermodynamics based on Information, World Scientific Pub. Co. Pte. Ltd. (2008). pp. 19–31.

Bodkin, R., Logic for All, Mercier Press, (1963). p. 21.

Brillouin, L., Science and Information Theory, 2nd Edn. New York: Academic Press. (1962).

Büchel, W., "Entropy and Information in the Universe", Nature (1967). 213; pp. 319–320.

Busse, H. & Hess, B., "Information transmission in a diffusion-coupled oscillatory chemical system", (1973). Nature, 244; pp. 203–205.

Campbell, B., "Biological Entropy Pump", Nature (1967). 215; p. 1038.

Carter, B., "Large Number Coincidences and the Anthropic Principle in Cosmology", In IAU Symposium 63: Confrontation of Cosmological Theories with Observational Data. Proceedings of the Symposium, Krakow, Poland, September 10-12, 1973. Dordrecht: Reidel. pp. 291–298. (Leslie, J., Ed.) (1974).

Cherry, C., On Human Communications, Cambridge Mass.: M.I.T. Press. (1957).

Cherry, C., "The Communication of Information", Endeavor (1964). 23; No. 88, pp. 13–17.

Clausius, R., "Über die bewegende Kraft der Wärme, Part I, Part II", Annalen der Physik (1850). 79; 368–397, 500–524. English Translation: "On the Moving Force of Heat, and the Laws regarding the Nature of Heat itself which are deducible therefrom." Phil. Mag. (1851). Fourth Series, 2; pp. 1–21, 102–119.

Conway, E.J., The Method of Isotopic Tracers Applied to the Study of

Active Transport, Pergamon Press, London, (1959). p. 111.

Davenport, H.W., "Substrate and Oxygen Consumption During Gastric Secretion", Fed. Proc. (1952). 11; pp. 715–721.

Davenport, H.W. and Chavré, L., "Acid Secretion and Oxygen Consumption In Vitro", Amer. J. Physiol. (1953). 174; pp. 203–208.

Denbigh, K., The Principles of Chemical Equilibrium, 3rd Edn. Cambridge: Cambridge University Press. (1971).

Eggers Jr., D.F., Gregory, N.W., Halsey, G.D. & Rabinovitch, B.S., Physical Chemistry, New York: Wiley. (1964). p. 105.

Fick, A., "Über Diffusion", Annalen der Physik und Chemie (Poggendorff), (1855). 94; pp. 59–86.

Gatlin, L.L., (1966). "The Information Content of DNA", J. Theor. Biol. 10; pp. 281–300.

Gatlin, L.L., (1968). "The Information Content of DNA. II.", J. Theor. Biol. 18; pp. 181–194.

Gatlin, L.L., Information Theory and the Living Organism. Columbia: Columbia University Press. (1972).

Goldman, S., Information Theory. New York: Prentice Hall. (1953).

Guilbaud, G.T., What is Cybernetics?, Heinemann, (1961).

Haldane, J.B.S., "The Origin of Life", Rationalist Annual (1928). 148; pp. 3–10.

Hargreaves, G. & Socrates, G., Elementary Chemical Thermodynamics, 3rd Edn. London: Butterworths, (1973). pp. 37-40.

Hartley, R.V.L., "Transmission of Information", Bell. Syst. Tech. J. (1928). 7; No. 3, July, pp. 535–563.

Hassenstein, B., Information and Control in the Living Organism. London: Chapman & Hall. (1971).

Holden, J.T., "Evolution of Transport Systems" (1968). J. Theor. Biol. 21; pp. 97–102.

Hoyle, F., The Nature of the Universe, Pelican Books, (1963). pp. 109–117.

Huxley, J., Evolution in Action, Pelican Books, (1963). pp. 12–15.

Khinchin, A.I., Mathematical Foundations of Information Theory. New York: Dover. (1957).

Koestler, A., The Act of Creation, Pan Books, London, (1966).

Lawden, D.F., "Chemical Evolution and the Origin of Life", Nature

(1964). 202; p. 412.

Lwoff, A., (1965). Biological Order. Cambridge, Mass.: M.I.T. Press.

MacKay, D.M., "Quantal Aspects of Scientific Information", Phil. Mag. (1950). 41; pp. 289–311.

Mandl, F., Statistical Physics. London: John Wiley & Sons. (1976).

Miller, G.A., "The Magical Number Seven, Plus-or-Minus Two: Some Limits on our Capacity for Processing Information", (1956). Psychol. Rev. 63; pp. 81–97.

Nash, L.K., Elements of Statistical Thermodynamics. London: Addison-Wesley Pub. Co. (1968).

O'Manique, J., Energy in Evolution: Teilhard's Physics of the Future, Garnstone Press, (1969). pp. 16–32.

Onsager, L., "Reciprocal Relations in Irreversible Processes. I.", (1931). Phys. Rev. 37; pp. 405–426.

Oparin, A.I., Life, its Nature, Origin and Development, Oliver & Boyd, (1961).

Paul, J., Cell Biology. London: Heinemann Studies in Biology. (1964).

Penrose, L.S., "Self-Reproducing Machines", Sci. Am. (1959). 200; June, pp. 105–114.

Ponnamperuma, C., "Chemical Evolution and the Origin of Life", Nature (1964). 201; pp. 337–340.

Popper, K.R., "The Arrow of Time", Nature (1956). 177; p. 538.

Popper, K.R., "Time's Arrow and Entropy", Nature (1965). 207; pp. 233–234.

Popper, K.R., "Time's Arrow and Feeding on Negentropy", Nature (1967a). 213; p. 320.

Popper, K.R., "Structural Information and the Arrow of Time", (1967b). Nature, 214; p. 322.

Ptitsyn, O.B. & Birshtein, T.M., "Method of determining the relative stability of different conformational states of biological macromolecules", Biopolymers (1969). 7; pp. 435–445.

Quastler, H., Emergence of Biological Organization. New Haven: Yale University Press. (1964).

Robertson, J., Economic Philosophy, Pelican Books, (1966). p. 8.

Ryan, J.P., "Information, Entropy and Various Systems", J. Theor. Biol. (1972). 36; pp. 139–146.

Ryan, J.P., "Aspects of Clausius-Shannon Identity: Emphasis on the Components of Transitive Information in Linear, Branched and Composite Physical Systems", Bull. Math. Biol. (1975). 37; pp. 223 – 254.

Ryan, J.P., "Information-Entropy Interfaces at Different Levels of Biological Organisation", J. Theor. Biol. (1980). 84; pp. 31–48.

Ryan, H., Ryan, J.P. and O'Connor, W.H., "The Effect of Diffusible Acids on Potassium Uptake by Yeast", Biochem. J. (1971). 125; pp. 1081–1086.

Ryan, J.P. and Ryan, H., "The Role of Intracellular pH in the Regulation of Cation Exchanges in Yeast", Biochem. J. (1972). 128; pp. 139–146.

Schrödinger, E., What is Life? The Physical Aspect of the Living Cell, (First published 1944). Based on lectures delivered under the auspices of the Dublin Institute for Advanced Studies at Trinity College, Dublin, in February 1943.

Shannon, C.E., The Mathematical Theory of Communication, University of Illinois Press, Urbana. (1949). cm.bell-labs.com/cm/ms/what/shannonday/shannon1948.pdf. Reprinted with corrections from Bell. Syst. Tech. J. (1948) 27; July, October, pp. 379–423, 623–656.

Sinnott, E.W., Matter, Mind and Man, Allen & Unwin, (1958).

Spanner, D.C., "The Active Transport of Water Under Temperature Gradients", Symp. Soc. Exp. Biol. (1954), VIII; pp. 76–93.

Szilard, L., "Über die Entropieverminderung in einem thermodynamischen System bei Eingriffen intelligenter Wesen" (On the Reduction of Entropy in a Thermodynamic System by the Interference of Intelligent Beings), Z. Physik, (1929). 53; pp. 840–856.

Teilhard de Chardin, P., The Appearance of Man, Harper, New York, (1965).

Teilhard de Chardin, P., The Future of Man, Collins, London, (1965).

Teilhard de Chardin, P., The Phenomenon of Man, Fontana Books, London, (1966).

Tribus, M. & McIrvine, C.E., "Energy and Information", Sci. Am. (1971). 225; pp. 179–188.

Viaud, G., Intelligence its Evolution and Forms, Arrow Books Ltd., Hutchinson & Co., (1960).

Viswanadham, C.R., "Entropy, Evolution and Living Systems", Nature (1968). 219; p. 653.

Warburton, F.E., "A Model of Natural Selection Based on a Theory of Guessing Games", J. Theor. Biol. (1967). 16; pp. 78–96.

Wheeler, J.A., "Law without Law", Quantum Theory and Measurement, (Wheeler, J.A. and Zurek W.H., Eds.), Princeton University Press, Princeton, NJ, USA. (1983). pp. 182–213.

Wheeler, J.A., "Information, Physics, Quantum: The Search for Links", Complexity, Entropy, and the Physics of Information. (Zurek, W.H., Ed.), Redwood City, California, Addison-Wesley. (1990). pp. 309–336.

Wilson, J.A., "Increasing Entropy of Biological Systems", Nature (1968a). 219; pp. 534–535.

Wilson, J.A., "Entropy, not Negentropy", Nature (1968b). 219; pp. 535–516.

Wisdom, J., Problems of Mind and Matter, Cambridge: Cambridge University Press, (1963).

Woolhouse, H.W., "Entropy and Evolution", Nature (1967). 216; p. 200.

Richard Rydon

About the Author

Richard Rydon is an award-winning science fiction novelist. His three books in the Luper Series, *The Oortian Summer* (2007), *The Omega Wave* (2008) and *The Palomar Paradox: A SETI Mystery* (2011), have been given excellent reviews. Richard's second novel, *The Omega Wave,* was selected as one of the finalists in the Science Fiction Category of the Reader Views Literary Awards and was awarded an Honorary Mention (Third Place) in the Reviewers Choice Awards in 2009.

Richard is an honours science graduate. More recently, he has obtained numerous certificates and diplomas in Psychology, Counselling, Theology, and a *Diplôme de Cuisine Française*. He is a prolific writer and has published over 300 papers, articles and poems, in scientific journals, magazines and local papers to date.

He has also published a second edition of his anthology of poetry, titled *A Golden Fuchsia-Laden Girl* (2011), containing 100 poems.

Books by Richard Rydon

Novels

The Luper Series

The Oortian Summer

The Omega Wave

The Palomar Paradox: A SETI Mystery

Poetry

A Golden Fuchsia-Laden Girl (Second Edition)

Short Play

A Children's Play

Non-Fiction

Matter, Energy and Mentality: Exploring Metaphysical Reality

About the Science Fiction Novels in the Luper Series

The Oortian Summer

'The Oortian Summer' is a romantic science fiction adventure involving co-worker relationships in an astronomical observatory as two massive comets approach the Earth. The unusual twist in the story involves a perilous attempt, proposed by Luper, the lead character, to bring the comets even closer to Earth to prevent a catastrophic geomagnetic flip.

The Omega Point

'The Omega Point' is a gothic science fiction novel. Aided and abetted by Quade their boss, Luper and Frieda progress secretly and meticulously, to develop biological computers called neurospheres. Working in the shadow of a rogue American Embassy, they first conceal but later reveal what they have seen and done.

The Palomar Paradox: A SETI Mystery

'The Palomar Paradox' sees Luper back in an astronomical observatory searching for signs of extraterrestrial intelligence. He finds himself working with Leila, a young girl recovering from leukaemia, and Karina, an experienced astronomer, among others. As their research continues, unusual signals are picked up by their radio telescope. The signals are explained, one by one, until ... !

About Richard Rydon's Poetry

A Golden Fuchsia-Laden Girl: 2nd Edition

'A Golden Fuchsia-Laden Girl: 2nd Edition' is an anthology of one hundred poems of whimsy, innocence and longing, by Richard Rydon, written and revised between 1957 and 2011. Twenty new poems have been added in this second edition.

www.ingramcontent.com/pod-product-compliance
Lightning Source LLC
Chambersburg PA
CBHW021908170526
45157CB00005B/2019